The National Interoperability  technical reference for emerg technicians responsible for ra The NIFOG includes rules and regulations for use of nationwide and other interoperability channels, tables of frequencies and standard channel names, and other reference material, formatted as a pocket-sized guide for radio technicians to carry with them.

If you are not familiar with interoperability and mutual aid communications, start with the "How to Use the National Interoperability Field Operations Guide" section.

We encourage you to program as many of these interoperability channels in your radios as possible, as permitted by the applicable regulations. Even if geographic restrictions on some channels preclude their use in your home area, you may have the opportunity to help in a distant location where the restrictions do not apply. Maximize your flexibility.

To download or request copies of the NIFOG, please visit

http://go.usa.gov/gTL

Your comments are welcome at NIFOG@HQ.DHS.GOV

Thank you.

Chris Essid, Director

Ross Merlin, NIFOG Author

DHS Office of Emergency Communications

TABLE OF CONTENTS

USING THE NATIONAL INTEROPERABILITY FIELD OPERATIONS GUIDE

What is the "National Interoperability Field Operations Guide"?

The "National Interoperability Field Operations Guide" (NIFOG) is a pocket-sized listing of land mobile radio (LMR) frequencies that are often used in disasters or other incidents where radio interoperability is required, and other information useful to emergency communicators.

Terms used in this document:

- FCC – Federal Communications Commission

- FCC Rules – contained in Title 47, Code of Federal Regulations (47CFR)

- Federal – used herein to differentiate between radio stations of the United States Government and those of any State, tribal, local, or regional governmental authority. "Federal Frequencies" refer to frequencies (channels) available for assignment to U.S. Government Agencies. Although the FCC is a Federal Government agency, the frequencies it administers are not "federal frequencies" - they are administered for state/tribal/local governments, commercial entities, and individuals.

- NCC - (1) the Public Safety National Coordination Committee, a Federal Advisory Committee formed by the FCC to advise it on interoperability; (2) National Coordinating Center for Telecommunications.

- NPSTC – the National Public Safety Telecommunications Council is a federation of organizations whose mission is to improve public safety communications and interoperability through collaborative leadership. After the charter for the NCC expired, NPSTC continued NCC's efforts to establish a common channel nomenclature. NPSTC channel IDs used in the NIFOG are based on the "Standard Channel Nomenclature for the Public Safety Interoperability Channels", APCO ANS 1.104.1-2010, approved June 9, 2010 by the American National Standards Institute (ANSI) - see http://www.npstc.org/documents/APCO-NPSTC-ANS1-104-1web.pdf

- NTIA – National Telecommunications and Information Administration

- NTIA Manual – The NTIA "Manual of Regulations and Procedures for Federal Radio Frequency Management" http://www.ntia.doc.gov/osmhome/redbook/redbook.html

- Radio frequencies are in MegaHertz (MHz) unless otherwise noted.

- CTCSS tone frequencies are in Hertz (Hz) or two-character Motorola codes.

- Emissions on frequencies above 30 MHz are narrowband analog FM, unless otherwise noted.

How is the NIFOG used?

The NIFOG may be used by radio technicians when programming channels in radios. We recommend having these channels programmed in radios at all times, as permitted by the applicable regulations, rather than waiting until a disaster is imminent or occurring to do the programming.

The NIFOG also is a useful tool for emergency communications planners, providing them with information on the interoperability channels most likely to be in the radios of responders from another discipline or jurisdiction.

Don't I need a license for these channels before programming them into radios?

If you are licensed under Part 90 of the FCC rules, you may program frequencies that you are not licensed to use IF "the communications involved relate directly to the imminent safety-of-life or property" or "with U.S. Government stations ... in connection with mutual activities" (see FCC rules 90.427 and 90.417).

However, note that 90.403(g) requires that "[f]or transmissions concerning the imminent safety-of-life or property, the transmissions shall be suspended as soon as the emergency is terminated." Also, the *safety of life* provision of 90.417(a) makes it clear that the exception applies only when the communications involved "relate directly" to the "imminent" safety of life or property. Because one overriding policy concern of the FCC is the prevention of harmful interference, any exceptions to the general prohibition on using non-licensed frequencies are limited in nature to responding to an imminent threat to safety-of-life or property, and licensees are not allowed to exceed the bounds of those communications.

See also 90.407 dealing with communications during an emergency which disrupts normal communications facilities and §90.411 dealing with civil defense communications.

There are no restrictions on U.S. Government stations programming frequencies into U.S. Government radios.

How can I use these frequencies if I don't have a license for them?

There are seven ways you can legally use these radio frequencies:

1. You or your employer may already have a Federal Communications Commission (FCC) license or a National Telecommunications and Information Administration (NTIA) authorization for some of the interoperability and mutual aid frequencies.

2. **For FCC licensees**, the non-Federal National Interoperability Channels VCALL10-VTAC14 and VTAC33-38, UCALL40-UCALL43D, and 8CALL90-8TAC94D are covered by a "blanket authorization" from the FCC - "Public safety licensees ... can operate mobile units on these interoperability channels without an individual license." See FCC 00-348, released 10/10/2000, paragraph 90.

3. You may operate on frequencies authorized to another licensee when that licensee designates you as a unit of their system, in accordance with FCC rule 90.421.

4. In extraordinary circumstances, the FCC may issue a "Special Temporary Authority" (STA) for such use in a particular geographic area.

5. In extraordinary circumstances, the NTIA may issue a "Temporary Assignment" for such use in a particular area.

6. **If you are an FCC Part 90 licensee**, you may operate a mobile station on the Federal Interoperability Channels only when authorized by the FCC (by license or STA) and only for interoperability with Federal radio stations authorized by the NTIA to use those channels. You **may not** use these channels for interoperability with other State,

tribal, regional, or local radio stations – these are not a substitute for your regular mutual aid channels. See FCC Public Notice DA 01-1621, released July 13, 2001.

7. When necessary for the IMMEDIATE protection of life or property, **FCC Part 90 licensees** may use prudent measures beyond the specifics of their license. See FCC rule 90.407, "Emergency communications". **U.S. Government stations** are authorized by NTIA rule 7.3.6 to operate on any Part 90 frequency with the permission of the FCC licensee when such use is necessary for communications directly related to the emergency at hand.

(FCC rules)

90.407 Emergency communications.

The licensee of any station authorized under this part may, during a period of emergency in which the normal communication facilities are disrupted as a result of hurricane, flood, earthquake or similar disaster, utilize such station for emergency communications in a manner other than that specified in the station authorization or in the rules and regulations governing the operation of such stations. The Commission may at any time order the discontinuance of such special use of the authorized facilities. [49 FR 36376, Sept. 17, 1984]

90.411 Civil defense communications.

The licensee of any station authorized under this part may, on a voluntary basis, transmit communications necessary for the implementation of civil defense activities assigned such station by local civil defense authorities during an actual or simulated emergency, including drills and tests. The Commission may at any time order the discontinuance of such special use of the authorized facilities. [49 FR 36376, Sept. 17, 1984]

(FCC rules - continued)
90.417 Interstation communication.

(a) Any station licensed under this part may communicate with any other station without restriction as to type, service, or licensee when the communications involved relate directly to the imminent safety-of-life or property.

(b) Any station licensed under this part may communicate with any other station licensed under this part, with U.S. Government stations, and with foreign stations, in connection with mutual activities, provided that where the communication involves foreign stations prior approval of the Commission must be obtained, and such communication must be permitted by the government that authorizes the foreign station. ...

90.421 Operation of mobile station units not under the control of the licensee.

Mobile stations, as defined in § 90.7, include vehicular-mounted and handheld units. Such units may be operated by persons other than the licensee ...

90.427 Precautions against unauthorized operation.

(a) ...

(b) Except for frequencies used in accordance with § 90.417, no person shall program into a transmitter frequencies for which the licensee using the transmitter is not authorized.

– 8 –

(NTIA rules)

7.3.4 Emergency Communications for which an Immediate Danger Exists to Human Life or Property

1. In situations where immediate danger exists to human life or property, an agency may operate temporarily on any regularly assigned frequency in a manner other than that specified in the terms of an existing assignment. Emergency operations under such situations should continue only as long as necessary to ensure that the danger to human life or property no longer exists. Emergency operations under these circumstances shall be reevaluated on a regular basis until such time as normal/ routine operations can be reestablished.

2. Interoperable communications for disaster/emergency response involving Federal, State, local, and tribal entities shall be in conformance with Section 4.3.16 of this Manual. Additional information regarding interoperable communications can also be found in the National Interoperability Field Operations Guide (NIFOG) ... promulgated by the Department of Homeland Security.

7.3.6 Emergency Use of Non-Federal Frequencies

In emergency situations, a government radio station may utilize any frequency authorized to a non-government radio station, under Part 90 of the FCC Rules and Regulations, when such use is necessary for communications with non-government stations and is directly related to the emergency at hand. Such use is subject to the following conditions:

a. The non-government licensee has given verbal or written concurrence.

b. Operations are conducted in accordance with the FCC Rules and Regulations.

c. Use is restricted to the service area and station authorization of the licensee.

d. All operations are under the direct control of the licensee and shall be immediately terminated when directed by the licensee.

e. Operations do not exceed 60 days.

f. A written report of each such use shall be provided, through the agency's FAS [Frequency Assignment Subcommittee, of NTIA's IRAC (Interdepartment Radio Advisory Committee)] representative, to the FCC as soon as practicable.

7.5.2 Frequencies Authorized by the FCC for Ship Stations

Frequencies authorized by the Federal Communications Commission for ship stations may be used by Government mobile stations to communicate with non-Government stations in the maritime mobile service.

7.5.3 Frequencies for the Safety of Life and Property

... (5) The frequency 40.5 MHz is designated as the military joint common frequency. Use of this channel is limited to communications necessary to establish contact when other channel information is not available and for emergency communications. This frequency also may be used for search and rescue communications.

(6) The provisions of this Manual do not prevent mobile stations, or mobile earth stations, in distress from using any frequency at its disposal to attract attention, make known its position, and obtain help. (See ITU Radio Regulation Ap. 13 Part A1, § 6,1.)

7.5.4 Frequencies for Coordinating Search and Rescue Operations

... (2) The frequency 123.1 MHz, using class A3E emission, may be used by stations of the aeronautical mobile service and by other mobile and land stations engaged in coordinated search and rescue operations.

(3) The frequency 156.3 MHz [VHF Marine channel 6] may be used for communications between ship stations and aircraft stations, using G3E emission, engaged in coordinated search and rescue (SAR) operations. When control of the scene of a SAR incident is under a Coast Guard coast station, 156.3 MHz may be used by ship stations to communicate with that coast station.

Does the NIFOG authorize me to use certain frequencies?

NO. The NIFOG does not grant authority to operate on any radio frequencies. Such authority can come only from the FCC or the NTIA.

Is the NIFOG the national emergency communications plan?

The NIFOG is the national guide for possible use in a situation where no other radio interoperability arrangement was promulgated by local authorities, or where emergency responders are unaware of such an arrangement. The NIFOG does NOT supersede any Federal, State, tribal, local, or regional emergency communications plan. If you are dispatched to a disaster or incident scene and have no other information on how to make contact with other emergency responders, the NIFOG provides useful suggestions for which frequencies to use to attempt initial contact.

Are the interoperability channels discussed in the NIFOG available nationwide?

No. Not all frequencies are available nationwide for use as described in the NIFOG. In particular, the "Non-Federal VHF Inland Interoperability Channels" may be used only in certain inland parts of the country, away from coastal areas and major waterways (see the map titled *Counties Where VTAC17/VTAC17D May Be Used* on page 28 for further details). Other channels in this plan may not be usable due to the potential for adjacent channel interference in some areas, or due to authorized on-channel uses that are different than the common uses described in the NIFOG.

For a detailed list of which counties are in which VHF Public Coast (VPC) area, see:
http://www.fcc.gov/oet/info/maps/areas/data/2000/FCCCNTY2K.txt and
http://www.fcc.gov/oet/info/maps/areas/data/2000/README_FCCCNTY2K.txt

FCC online area cross-reference search: http://www.fcc.gov/fcc-bin/cesearch.pl

Who do I contact to use interoperability channels?

These channels can be used where licensed or authorized by FCC or NTIA, including authorization by a STA. The COML (Communications Unit Leader) acts as, or delegates the role of frequency manager; assigning specific uses to available radio channels and coordinating with the FCC and NTIA for authorization to use additional channels as needed.

If access to the COML has not been pre-arranged or is not working as planned, try the calling channels specified in the NIFOG at or near the command post, incident scene, or staging area.

At a Federally-declared disaster where a Joint Field Office (JFO) is established, "Communications" is under the Logistics Section and the Operations Section - doesn't that contradict ICS?

No. Communications for the personnel working in the JFO is the responsibility of the JFO Communications Unit, which is under the Logistics Section. Communications for those affected by the disaster, including local first responders, victims, and local infrastructure, as well as Federal assets supporting local disaster operations, is the focus of the Disaster Emergency Communications (DEC) Branch (ESF #2 - Communications), which is in the JFO Operations Section. The DEC Branch may have personnel in the Tactical Communications Group, Wireless Communications Task Force filling the role of Spectrum Manager. Working with the Logistics Section Communications Unit Leader and the local COML responsible for the affected area, Wireless Communications Task Force Leader coordinates the use of radio frequencies used by Federal responders with State and local government authorities. The Wireless Communications Task Force provides direct access to the FCC and NTIA decision-makers.

How do I request a Special Temporary Authorization (STA)?

FCC licensees request a Special Temporary Authorization (STA) from the FCC:

During Normal FCC Business Hours (Monday through Friday, 8:00am - 5:30pm EST/EDT)
Tracy Simmons - STA Licensing (Part 90—Land Mobile and Public Safety), Public Safety & Homeland
Security Bureau - phone: 717-338-2657   email: Tracy.Simmons@fcc.gov
or file electronically: FCC Form 601 - ULS   http://wireless.fcc.gov/uls/ then click on "ULS Online Filing"

Outside of Normal FCC Business Hours (5:30pm - 8am EST/EDT, weekends, and holidays)
Communications and Crisis Management Center (CCMC) -
phone: 202-418-1122   email: disasterassistance@fcc.gov

First Responders and Public Safety Entities with general STA inquiries

Allan Manuel, Public Safety & Homeland Security Bureau
phone: 202-418-1164 mobile: 202-391-5331   email: Allan.Manuel@fcc.gov

or

FCC 24/7 Operations Center   phone: 202-418-1122   email: FCCOPCenter@fcc.gov

U.S. Government radio stations request temporary assignment or STAs via their agency representative to the Frequency Assignment Subcommittee (FAS) of the Interdepartment Radio Advisory Committee (IRAC). The telephone number for the NTIA Frequency Assignment Branch is 202-482-1132.
[See the previous page for requesting STAs when a Joint Field Office is operational for an incident.]

Does the NIFOG specify exactly how to program channels?

No. There is no one-size-fits-all solution due to differing radio designs. The NIFOG uses the ANSI "Standard Channel Nomenclature for the Public Safety Interoperability Channels" for channel names - see "NPSTC" on page 2.

For some channels, the standard nomenclature specifies a "direct" ("talk-around") channel for repeaters which takes up an additional memory slot. Some radios have a switch that permits talk-around on a repeater channel, and using this feature would save memory slots. Similarly, some radios may have a switch or button to enable or disable receive CTCSS; for radios that don't, another channel may be programmed so both modes would be available.

Until the narrowband transition is complete, some mutual aid channels may be wideband in some areas and narrowband in others. The standard nomenclature does not always address how to label the same frequencies with different bandwidths. For the legacy police, EMS, and fire mutual aid channels 155.475, 155.340, 154.265, 154.280 and 154.295, we suggest VLAW31W, VMED28W, VFIRE22W, VFIRE21W, and VFIRE23W as the wideband names for VLAW31, VMED28, VFIRE22, VFIRE21, and VFIRE23 on the same frequencies. For the SAR common channel, 155.160 MHz, we suggest "SAR WFM" for wideband and "SAR NFM" for narrowband.

Also, consider programming additional VHF Marine channels as possible interoperability channels (for use when properly authorized), based on local or regional use. In particular, channels used by drawbridge tenders may be appropriate; see http://wireless.fcc.gov/marine/vhfchan1.pdf for authorized channel uses and http://www.navcen.uscg.gov/?pageName=mtVhf for frequencies.

Recommended modes for using Federal Interoperability Channels: use analog for all Incident Response channels (CTCSS 167.9 Hz) and Law Enforcement channels LE A, LE 1, LE B, LE 10, and LE 16 (CTCSS 167.9 Hz); use P25 digital for the remaining LE channels, NAC $68F. CTCSS should always be transmitted on the analog channels, but carrier squelch (CSQ, no CTCSS) should be used on receive. Consider allowing the user to enable or disable CTCSS on receive by a switch or button; otherwise use CSQ on receive.

Should Fire/EMS radios have the Law Enforcement interoperability channels programmed, and vice versa?

Yes. Radios for public safety personnel should have as many of these interoperability channels programmed as possible, as permitted by the applicable regulations. Interoperability may require crossing jurisdictional and functional lines. On the Federal interoperability channels, "Incident Response" (IR) means everybody – Fire, Rescue, EMS, Public Works, Law Enforcement, etc. The "Law Enforcement" (LE) channels will be used "primarily" for Law Enforcement activities, but could be designated for other incident support if that would not hamper Law Enforcement activities, and if assigned by the agency in control of the incident.

How do emergency responders use the calling channels?

As you approach an incident scene or staging area, you might establish contact on a dispatch or working channel. If you can't make contact, or if no channel was designated for this purpose, attempt to make contact on one of the designated interoperability calling channels. If it is a repeater channel and you get no response, try the "direct" or "talk-around" mode if your radio has that capability. In some cases, the talk-around channel exists as a distinct channel on the radio. For example, the VHF Incident Response Federal Interoperability Channel is known as "NC 1". The talk-around for this repeater channel is known as "IR 5".

The non-Federal national interoperability calling channels are VCALL10, UCALL40, and 8CALL90; the Federal IR and LE calling channels are "NC 1" (direct: "IR 5"), "NC 2" (direct: "IR 15"), "LE A", and "LE B". If you are unable to make contact on these channels, consider the wideband interoperability channels – if you are authorized to use them, or if your situation qualifies as "IMMEDIATE protection of life or property". You may be able to learn what you need without transmitting, by just listening to radio traffic on one of these channels.

How do Search and Rescue personnel on land, on watercraft, and on aircraft coordinate by radio?

Certain VHF Marine channels are designated in this plan for Search and Rescue (SAR) interoperability. Searchers on land, in boats, and in aircraft need to be able to communicate with each other to coordinate rescues. There is no VHF channel authorized and readily available to all three communities. Some aircraft involved in SAR have VHF Marine radios, as do most boaters; the VHF radios that many ground SAR groups use are capable of covering the VHF Marine frequencies. We recommend that SAR participants have the channels in this plan pre-programmed in their radios. VHF Marine channels shall not be used for conventional, terrestrial search and rescue operations – they are in this plan due to the likelihood of boats being involved in SAR in coastal areas. Also, 155.16 MHz is licensed to many SAR organizations. We encourage public safety entities to obtain licenses for this frequency to facilitate interoperability. Likewise, we encourage SAR organizations with VHF radios to program the appropriate VHF Marine channels in their radios and to exercise great restraint in using these channels only when authorized.

How can I get answers to questions about the "National Interoperability Field Operations Guide", or how can I offer suggestions to improve it?

Please send your questions or comments to the U.S. Department of Homeland Security, Office of Emergency Communications, at OEC@HQ.DHS.GOV and include your name, agency or organization affiliation, and your e-mail address.

How do I get copies of the NIFOG?

The latest version of the NIFOG can be downloaded or ordered from http://go.usa.gov/gTL

Recommendations for Programming the Federal Interoperability Channels

1. If there is enough room in your radio, program all channels as analog and again as digital channels. If not, program as follows:

   a. Incident Response channels – all analog.

   b. Law Enforcement channels – program all as P25 digital with NAC $68F except LE A, LE 1, LE B, LE10, and LE 16 which are to be programmed analog with Tx CTCSS 167.9 Hz (6Z) and no Rx CTCSS (carrier squelch, CSQ)

2. If your radio has a user-selectable option to enable/disable CTCSS on receive, you may choose to configure this option so that the user can enable the same CTCSS tone used on transmit for receive. The default configuration should be CSQ receive.

Note on using the Federal Interoperability Channels: These channels may not be used for state/state, state/local, or local/local interoperability. A Federal entity must be involved when these are used.

Regulations and Guidelines for National Interoperability

1. The FCC and NTIA rules allow for some flexibility in frequency use by personnel directly involved in a situation where there is imminent danger to human life or property. This does NOT mean "in an emergency, anything goes."

2. For communications not covered by #1, your use of a radio frequency must be authorized by:

a. Your (or your agency's) FCC license or NTIA authorization

b. "License by rule" – a provision in FCC rules that authorizes use of a radio frequency under specified conditions without a specific license or authorization issued to the user

c. A "Special Temporary Authorization" provided by FCC or NTIA

3. Digital P25 operations on non-Federal interoperability channels should transmit the default Network Access Code (NAC) $293, and receive with NAC $F7E (accept any incoming NAC). Specify talkgroup $FFFF, which includes everyone.

4. Default modes for using Federal Interoperability Channels: use analog for all Incident Response channels and Law Enforcement channels LE A, LE 1, LE B, LE 10, and LE 16; use P25 Digital for the remaining LE channels, NAC $68F.

Conditions for Use of Federal Interoperability Channels

1. The "VHF Incident Response (IR) Federal Interoperability Channel Plan", the "UHF Incident Response (IR) Federal Interoperability Channel Plan", the "VHF Law Enforcement (LE) Federal Interoperability Channel Plan", and the "UHF Law Enforcement (LE) Federal Interoperability Channel Plan" show frequencies available for use by all Federal agencies to satisfy law enforcement and public safety incident response interoperability requirements. These frequencies will be referred to hereinafter as "Federal Interoperability Channels".

2. The Federal Interoperability Channels are available for use among Federal agencies and between Federal agencies and non-federal entities with which Federal agencies have a requirement to operate.

3. The channels are available to non-federal entities to enable joint Federal/non-federal operations for law enforcement and incident response, subject to the condition that harmful interference will not be caused to Federal stations. These channels are restricted to interoperability communications and are not authorized for routine or administrative uses.

4. Extended operations and congestion may lead to frequency conflicts. Coordination with NTIA is required to resolve these conflicts.

5. Only narrowband emissions are to be used on the Federal Interoperability Channels.

6. Equipment used (transmitters and receivers) must meet the standards established in Section 5.3.5.2 of the NTIA Manual:

   a. TIA/EIA 603-B for narrowband analog;

   b. TIA TSB 102.CAAB-A for narrowband digital

7. A complete listing of conditions for use by Federal users can be found in Section 4.3.16 of the NTIA Manual.

8. Use of these frequencies within 75 miles of the Canadian border and 5 miles of the Mexican border require special coordination and in some cases will not be available for use.

Law Enforcement Plans

1. Frequencies 167.0875 MHz and 414.0375 MHz are designated as National Calling Channels for initial contact and will be identified in the radio as indicated in the Law Enforcement Federal Interoperability Channel Plans.

2. Initial contact communications will be established using narrowband analog FM emission (11K25F3E).

3. The interoperability channels will be identified in mobile and portable radios as indicated in the Law Enforcement Federal Interoperability Channel Plans with Continuous Tone-Controlled Squelch Systems (CTCSS) frequency 167.9 Hz and/or Network Access Code (NAC) $68F.

Incident Response Plans

1. Frequencies 169.5375 MHz (paired with 164.7125 MHz) and 410.2375 MHz (paired with 419.2375 MHz) are designated as the calling channels for initial contact and will be identified in the radio as indicated in the Incident Response Federal Interoperability Channel Plans.

2. Initial contact will be established using narrowband analog FM emission (11K25F3E).

3. To ensure access by stations from outside the normal area of operation, Continuous Tone-Controlled Squelch Systems (CTCSS) will not be used on the calling channels.

4. The interoperability channels will be identified in mobile and portable radios as indicated in the "VHF Incident Response (IR) Federal Interoperability Channel Plan" and the "UHF Incident Response (IR) Federal Interoperability Channel Plan".

FCC Rules and Regulations

Title 47, Code of Federal Regulations, Parts 0-199

http://wireless.fcc.gov/rules.html

Part 80	Maritime Services
	For information on VHF Marine channels, see
	http://www.navcen.uscg.gov/?pageName=mtVhf
Part 87	Aviation Services
Part 90	Private Land Mobile Radio Services
Part 95	Personal Radio Services (includes GMRS, FRS, CB, & MURS)
Part 97	Amateur Radio Service

NTIA Rules and Regulations

Title 47, Code of Federal Regulations, Part 300

http://www.ntia.doc.gov/osmhome/redbook/redbook.html

INTEROPERABILITY CHANNELS

Non-Federal VHF National Interoperability Channels

Description	Channel Name	Mobile Receive Frequency	Mobile Transmit Frequency	CTCSS Tone ±
VHF Low Band				
Law Enforcement	LLAW1	39.4600	45.8600	CSQ /156.7 (5A)
	LLAW1D	39.4600	39.4600	CSQ /156.7 (5A)
Fire (Proposed)	LFIRE2	39.4800	45.8800	CSQ /156.7 (5A)
	LFIRE2D	39.4800	39.4800	CSQ /156.7 (5A)
Law Enforcement	LLAW3	45.8600	39.4600	CSQ /156.7 (5A)
	LLAW3D	45.8600	45.8600	CSQ /156.7 (5A)
Fire (Proposed)	LFIRE4	45.8800	39.4800	CSQ /156.7 (5A)
Fire	LFIRE4D	45.8800	45.8800	CSQ /156.7 (5A)

Frequency 39.4800 MHz is pending FCC assignment for exclusive fire intersystem use.

± Default operation should be carrier squelch receive, CTCSS transmit. If the user can enable/disable without reprogramming the radio, the indicated CTCSS tone also could be programmed for receive, and the user instructed how and when to enable/disable.

Non-Federal VHF National Interoperability Channels

VHF High Band

Description	Channel Name	Mobile Receive Freq.	Mobile Transmit Freq.	CTCSS Tone
Calling	VCALL10	155.7525	155.7525	CSQ / 156.7 (5A) ±
Tactical	VTAC11 *	151.1375	151.1375	CSQ / 156.7 (5A) ±
Tactical	VTAC12 *	154.4525	154.4525	CSQ / 156.7 (5A) ±
Tactical	VTAC13	158.7375	158.7375	CSQ / 156.7 (5A) ±
Tactical	VTAC14	159.4725	159.4725	CSQ / 156.7 (5A) ±
Tac Rpt	VTAC33 * •	159.4725	151.1375	CSQ / 136.5 (4Z)
Tac Rpt	VTAC34 * •	158.7375	154.4525	CSQ / 136.5 (4Z)
Tac Rpt	VTAC35 •	159.4725	158.7375	CSQ / 136.5 (4Z)
Tac Rpt	VTAC36 * •	151.1375	159.4725	CSQ / 136.5 (4Z)
Tac Rpt	VTAC37 * •	154.4525	158.7375	CSQ / 136.5 (4Z)
Tac Rpt	VTAC38 •	158.7375	159.4725	CSQ / 136.5 (4Z)

*VTAC11-12, VTAC33-34, and VTAC36-37 may not be used in Puerto Rico or the USVI.

±Default operation should be carrier squelch receive, CTCSS transmit. If the user can enable/disable without re-programming the radio, the indicated CTCSS tone also could be programmed for receive, and the user instructed how and when to enable/disable.

• VTAC33-38 recommended for deployable tactical repeater use only (FCC Station Class FB2T).

• VTAC33-35 are preferred; VTAC33-35 should be used only when necessary due to interference.

All channels on this page are NARROWBAND only.

Non-Federal VHF National Interoperability Channels			
VHF Inland			
Description	Channel Name	Mobile RX (MHz)	Mobile TX (MHz)
Tactical – narrowband FM	VTAC17	161.8500	157.2500
Tactical – narrowband FM	VTAC17D	161.8500	161.8500

Default operation should be carrier squelch receive, CTCSS 156.7 Hz(5A) transmit. If the user can enable/disable CTCSS without reprogramming the radio, the indicated CTCSS tone also could be programmed for receive, and the user instructed how and when to enable/disable.

For VTAC17/VTAC17D only: Base stations: 50 watts max, antenna HAAT 400 feet max. Mobile stations: 20 watts max, antenna HAAT 15 feet max. These channels are for tactical use and may not be operated on board aircraft in flight. These channels use narrowband FM and are available only in certain inland areas at least 100 miles from a major waterway. These channels use the same frequencies as VHF Marine channel 25, which uses wideband FM. Use only where authorized. See map on next page. In these authorized areas, interoperability communications have priority over grandfathered public coast and public safety licensees.

All channels on this page are NARROWBAND only.

Counties Where VTAC17/VTAC17D May Be Used

Numbers Indicate VHF Public Coast Station Areas - see 47CFR80.371(c)(ii)

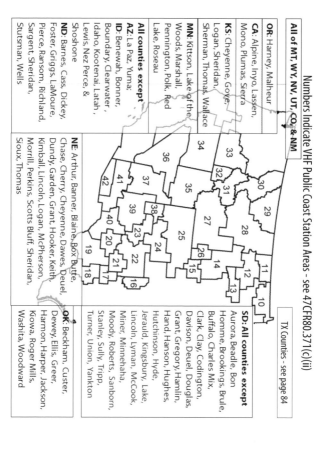

All of MT, WY, NV, UT, CO & NM

OR: Harney, Malheur

CA: Alpine, Inyo, Lassen, Mono, Plumas, Sierra

KS: Cheyenne, Gove, Logan, Sheridan, Sherman, Thomas, Wallace

MN: Kittson, Lake of the Woods, Marshall, Pennington, Polk, Red Lake, Roseau

All counties except:
AZ: La Paz, Yuma;
ID: Benewah, Bonner, Boundary, Clearwater, Idaho, Kootenai, Latah, Lewis, Nez Perce, & Shoshone
ND: Barnes, Cass, Dickey, Foster, Griggs, LaMoure, Pierce, Ransom, Richland, Sargent, Sheridan, Stutsman, Wells

NE: Arthur, Banner, Blaine, Box Butte, Chase, Cherry, Cheyenne, Dawes, Deuel, Dundy, Garden, Grant, Hooker, Keith, Kimball, Lincoln, Logan, McPherson, Morrill, Perkins, Scotts Bluff, Sheridan, Sioux, Thomas

TX Counties - see page 84

SD: All counties except
Aurora, Beadle, Bon Homme, Brookings, Brule, Buffalo, Charles Mix, Clark, Clay, Codington, Davison, Deuel, Douglas, Grant, Gregory, Hamlin, Hand, Hanson, Hughes, Hutchinson, Hyde, Jerauld, Kingsbury, Lake, Lincoln, Lyman, McCook, Miner, Minnehaha, Moody, Roberts, Sanborn, Stanley, Sully, Tripp, Turner, Union, Yankton

OK: Beckham, Custer, Dewey, Ellis, Greer, Harmon, Harper, Jackson, Kiowa, Roger Mills, Washita, Woodward

Non-Federal UHF National Interoperability Repeater Channels

Description	Channel Name	Mobile RX (MHz)	Mobile TX (MHz)
Calling	UCALL40	453.2125	458.2125
Calling	UCALL40D	453.2125	453.2125
Tactical	UTAC41	453.4625	458.4625
Tactical	UTAC41D	453.4625	453.4625
Tactical	UTAC42	453.7125	458.7125
Tactical	UTAC42D	453.7125	453.7125
Tactical	UTAC43	453.8625	458.8625
Tactical	UTAC43D	453.8625	453.8625

Default operation should be carrier squelch receive, CTCSS 156.7(5A) transmit. If the user can enable/disable CTCSS without reprogramming the radio, the indicated CTCSS tone also could be programmed for receive, and the user instructed how and when to enable/disable. All channels on this page are NARROWBAND only.

700 MHz Interoperability Channels				
FCC Channel (Subscriber Load)		Transmit and Receive Frequencies	Primary Use	Channel Name
Receive Ch.	Transmit Ch.			
23-24	983-984	799.14375	General	7TAC51
	23-24	769.14375	Public Safety	7TAC51D
39-40	999-1000	799.24375	Calling	7CALL50
	39-40	769.24375	Channel	7CALL50D
63-64	1023-1024	799.39375	EMS	7MED65
	63-64	769.39375		7MED65D
79-80	1039-1040	799.49375	EMS	7MED66
	79-80	769.49375		7MED66D
103-104	1063-1064	799.64375	General	7TAC52
	103-104	769.64375	Public Safety	7TAC52D
119-120	1079-1080	799.74375	General	7TAC55
	119-120	769.74375	Public Safety	7TAC55D
143-144	1103-1104	799.89375	Fire	7FIRE63
	143-144	769.89375		7FIRE63D
159-160	1119-1120	799.99375	Fire	7FIRE64
	159-160	769.99375		7FIRE64D
183-184	1143-1144	800.14375	General	7TAC53
	183-184	770.14375	Public Safety	7TAC53D
199-200	1159-1160	800.24375	General	7TAC56
	199-200	770.24375	Public Safety	7TAC56D
223-224	1183-1184	800.39375	Law	7LAW61
	223-224	770.39375	Enforcement	7LAW61D

FCC Channel (Subscriber Load)		Transmit and Receive Frequencies	Primary Use	Channel Name
Receive Ch.	Transmit Ch.			
239-240	1199-1200	800.49375	Law	7LAW62
	239-240	770.49375	Enforcement	7LAW62D
263-264	1223-1224	800.64375	General	7TAC54
	263-264	770.64375	Public Safety	7TAC54D
279-280	1239-1240	800.74375	Mobile Data	7DATA69
	279-280	770.74375		7DATA69D
303-304	1263-1264	800.89375	Mobile	7MOB59
	303-304	770.89375	Repeater	7MOB59D
319-320	1279-1280	800.99375	Other Public	7GTAC57
	319-320	770.99375	Service	7GTAC57D
641-642	1601-1602	803.00625	EMS	7MED86
	641-642	773.00625		7MED86D
657-658	1617-1618	803.10625	General	7TAC71
	657-658	773.10625	Public Safety	7TAC71D
681-682	1641-1642	803.25625	Calling	7CALL70
	681-682	773.25625	Channel	7CALL70D
697-698	1657-1658	803.35625	EMS	7MED87
	697-698	773.35625		7MED87D
721-722	1681-1682	803.50625	Fire	7FIRE83
	721-722	773.50625		7FIRE83D
737-738	1697-1698	803.60625	General	7TAC72
	737-738	773.60625	Public Safety	7TAC72D

FCC Channel (Subscriber Load)		Transmit and Receive Frequencies	Primary Use	Channel Name
Receive Ch.	Transmit Ch.			
761-762	1721-1722	803.75625	General	7TAC75
	761-762	773.75625	Public Safety	7TAC75D
777-778	1737-1738	803.85625	Fire	7FIRE84
	777-778	773.85625		7FIRE84D
801-802	1761-1762	804.00625	Law	7LAW81
	801-802	774.00625	Enforcement	7LAW81D
817-818	1777-1778	804.10625	General	7TAC73
	817-818	774.10625	Public Safety	7TAC73D
841-842	1801-1802	804.25625	General	7TAC76
	841-842	774.25625	Public Safety	7TAC76D
857-858	1817-1818	804.35625	Law	7LAW82
	857-858	774.35625	Enforcement	7LAW82D
881-882	1841-1842	804.50625	Mobile	7MOB79
	881-882	774.50625	Repeater	7MOB79D
897-898	1857-1858	804.60625	General	7TAC74
	897-898	774.60625	Public Safety	7TAC74D
921-922	1881-1882	804.75625	Mobile Data	7DATA89
	921-922	774.75625		7DATA89D
937-938	1897-1898	804.85625	Other Public	7GTAC77
	937-938	774.85625	Service	7GTAC77D

12.5 kHz narrowband channels shown as odd-even channel pairs of 6.25 kHz channels.
Ref: http://www.apco911.org/frequency/documents/700_NB_channel_centers.pdf

Non-Federal 800 MHz National Mutual Aid Repeater Channels

Description	Ch. Name	Mobile RX (MHz)*	Mobile TX (MHz)*
Calling	8CALL90	851.0125 (866.0125)	806.0125 (821.0125)
Calling – Direct	8CALL90D	851.0125 (866.0125)	851.0125 (866.0125)
Tactical	8TAC91	851.5125 (866.5125)	806.5125 (821.5125)
Tactical – Direct	8TAC91D	851.5125 (866.5125)	851.5125 (866.5125)
Tactical	8TAC92	852.0125 (867.0125)	807.0125 (822.0125)
Tactical – Direct	8TAC92D	852.0125 (867.0125)	852.0125 (867.0125)
Tactical	8TAC93	852.5125 (867.5125)	807.5125 (822.5125)
Tactical – Direct	8TAC93D	852.5125 (867.5125)	852.5125 (867.5125)
Tactical	8TAC94	853.0125 (868.0125)	808.0125 (823.0125)
Tactical – Direct	8TAC94D	853.0125 (868.0125)	853.0125 (868.0125)

Default operation should be carrier squelch receive, CTCSS 156.7(5A) transmit. If the user can enable/disable CTCSS without reprogramming the radio, the indicated CTCSS tone could also be programmed for receive, and the user instructed how and when to enable/disable.

*The frequency in parenthesis, which is 15 MHz higher, is the frequency used before rebanding - channel names were ICALL, ITAC1 - ITAC4. Wideband FM 20K0F3E before and after rebanding.

VHF Incident Response (IR) Federal Interoperability Channel Plan

Suggested Assignment (subject to availability & local plans)	Channel Name	Note	Mobile RX (MHz)	Mobile TX (MHz)
Incident Calling	NC 1	Calling	169.5375	164.7125
Incident Command	IR 1		170.0125	165.2500
Medical Evacuation Control	IR 2		170.4125	165.9625
Logistics Control	IR 3		170.6875	166.5750
Interagency Convoy	IR 4		173.0375	167.3250
Incident Calling (Direct)	IR 5	Direct for NC 1 Calling	169.5375	169.5375 (S)
Incident Command (Direct)	IR 6	Direct for IR 1	170.0125	170.0125 (S)
Medical Evacuation Control (Direct)	IR 7	Direct for IR 2	170.4125	170.4125 (S)
Logistics Control (Direct)	IR 8	Direct for IR 3	170.6875	170.6875 (S)
Interagency Convoy (Direct)	IR 9	Direct for IR 4	173.0375	173.0375 (S)

*See "Conditions for Use of Federal Interoperability Channels" on pages 22 - 24.
Default operation should be carrier squelch receive, CTCSS 167.9/CSQ transmit. If the user can enable/disable CTCSS without reprogramming the radio, the indicated CTCSS tone also could be programmed for receive, and the user instructed how and when to enable/disable.
All channels on this page are NARROWBAND only.

UHF Incident Response (IR) Federal Interoperability Channel Plan

Suggested Assignment (subject to availability & local plans)	Channel Name	Note	Mobile RX (MHz)	Mobile TX (MHz)
Incident Calling	NC 2	Calling	410.2375	419.2375
Ad hoc assignment	IR 10		410.4375	419.4375
Ad hoc assignment	IR 11		410.6375	419.6375
SAR Incident Command	IR 12		410.8375	419.8375
Ad hoc assignment	IR 13		413.1875	413.1875 (S)
Interagency Convoy	IR 14		413.2125	413.2125 (S)
Incident Calling (Direct)	IR 15	Direct for NC 2 Calling	410.2375	410.2375 (S)
Ad hoc assignment (Direct)	IR 16	Direct for IR 10	410.4375	410.4375 (S)
Ad hoc assignment (Direct)	IR 17	Direct for IR 11	410.6375	410.6375 (S)
SAR Incident Command (Direct)	IR 18	Direct for IR 12	410.8375	410.8375 (S)

*See "Conditions for Use of Federal Interoperability Channels" on pages 22 - 24.
Default operation should be carrier squelch receive, CTCSS 167.9/CSQ transmit. If the user can enable/disable CTCSS without reprogramming the radio, the indicated CTCSS tone also could be programmed for receive, and the user instructed how and when to enable/disable.
All channels on this page are NARROWBAND only.

VHF Law Enforcement (LE) Federal Interoperability Channel Plan

Description	Channel Name	Note	Mobile RX (MHz)	Mobile TX (MHz)	CTCSS or NAC
Calling	LE A	Analog	167.0875	167.0875 (S)	167.9 Tx, CSQ Rx
Tactical	LE 1	Analog	167.0875	162.0875	167.9 Tx, CSQ Rx
Tactical	LE 2		167.2500	162.2625	$68F
Tactical	LE 3		167.7500	162.8375	$68F
Tactical	LE 4		168.1125	163.2875	$68F
Tactical	LE 5		168.4625	163.4250	$68F
Tactical	LE 6	Direct for LE 2	167.2500	167.2500 (S)	$68F
Tactical	LE 7	Direct for LE 3	167.7500	167.7500 (S)	$68F
Tactical	LE 8	Direct for LE 4	168.1125	168.1125 (S)	$68F
Tactical	LE 9	Direct for LE 5	168.4625	168.4625 (S)	$68F

*See "Conditions for Use of Federal Interoperability Channels" on pages 22 - 24.
CTCSS on receive only if user selectable; else CSQ.
All channels on this page are NARROWBAND only.

UHF Law Enforcement (LE) Federal Interoperability Channel Plan

Description	Channel Name	Note	Mobile RX (MHz)	Mobile TX (MHz)	CTCSS or NAC
Calling	LE B	Analog	414.0375	414.0375 (S)	167.9 Tx, CSQ Rx
Tactical	LE 10	Analog	409.9875	418.9875	167.9 Tx, CSQ Rx
Tactical	LE 11		410.1875	419.1875	$68F
Tactical	LE 12		410.6125	419.6125	$68F
Tactical	LE 13		414.0625	414.0625 (S)	$68F
Tactical	LE 14		414.3125	414.3125 (S)	$68F
Tactical	LE 15		414.3375	414.3375 (S)	$68F
Tactical	LE 16	Direct for LE 10 - Analog	409.9875	409.9875 (S)	167.9 Tx, CSQ Rx
Tactical	LE 17	Direct for LE 11	410.1875	410.1875 (S)	$68F
Tactical	LE 18	Direct for LE 12	410.6125	410.6125 (S)	$68F

*See "Conditions for Use of Federal Interoperability Channels" on pages 22 - 24. CTCSS on receive only if user selectable; else CSQ. All channels on this page are NARROWBAND only.

Federal / Non-Federal SAR Command Interoperability Plan			
Channel Name*	Mobile RX (MHz)	Mobile TX (MHz)	CTCSS
IR 12**	410.8375	419.8375	167.9 Tx, CSQ Rx
VTAC14	159.4725	159.4725	156.7 Tx, CSQ Rx (156.7 Rx if user selectable)
UTAC43	453.8625	458.8625	156.7 Tx, CSQ Rx (156.7 Rx if user selectable)
8TAC94 (ITAC4 before rebanding)	853.0125 (868.0125 before rebanding)	808.0125 (823.0125 before rebanding)	156.7 Tx, CSQ Rx (156.7 Rx if user selectable)
VHF Marine Ch. 17***	156.8500 (this use requires FCC STA)	156.8500 (this use requires FCC STA)	none

* If a repeater is not available, substitute the corresponding talk-around channel: IR 18 for IR 12, UTAC43D for UTAC43, 8TAC94D for 8TAC94.

** See Conditions for Use of Federal Interoperability Channels on pages 22 - 24.

*** VHF marine ch. 17 is wideband FM, emission 16K0F3E.

– 38 –

Federal / Non-Federal VHF SAR Operations Interoperability Plan

Suggested SAR Function	Frequency (MHz)
Ground Operations	155.1600 narrowband FM (or wideband FM till 1/1/2013)
Maritime Operations *	157.050 or 157.150 (VHF Marine ch.21A or 23A) as specified by USCG Sector Commander
Air Operations – civilian	123.100 MHz AM (may not be used for tests or exercises)
Air Operations – USCG/Military	345.0 MHz AM for initial contact only; then move to 282.8 MHz AM or other working channel
Air rescue assets to air rescue assets (deconfliction)	As charted on standard air chart or MULTICOM 122.850 (south or west sector) & 122.900 MHz (north or east sector), or as specified by FAA. 122.850 may not be used for tests or exercises
Ground to Air SAR working channel	157.175 83A (21A, 23A, 81A alternates as specified by local USCG Sector Commander) **
Ground to Maritime SAR working channel	157.050 21A (23A, 81A, 83A alternates as specified by local USCG Sector Commander) **
Maritime/Air/Ground SAR working channel *	157.175 83A (21A, 23A, 81A alternates as specified by local USCG Sector Commander) **
EMS / Medical Support	155.3400 narrowband FM (or wideband FM till 1/1/2013)
Hailing* & DISTRESS only - Maritime/Air/Ground	156.800 VHF Marine channel 16 *

* Use VHF Marine ch.16 to make contact (30 seconds max.), then move to appropriate working channel as directed by local USCG Sector Commander. Non-maritime use of any VHF Marine channel requires FCC Special Temporary Authority or appropriate license. VHF marine channels use wideband FM, emission 16K0F3E

** VHF Marine channels: 16=156.800 21A=157.050 22A=157.100 23A=157.150 81A=157.075 82A=157.125 83A=157.1750

Direction from USCG, FCC, or FAA overrides information in this table. This table does not convey authority to operate.

– 39 –

VHF Public Safety Mutual Aid and Common Channels

Frequency (MHz)	Usage	Wideband ID	Narrowband ID	Note
155.1600	Search and Rescue Common	SAR WFM	SAR NFM	Not designated by FCC; availability varies.
154.2650 mobile	Fire Mutual Aid	VFIRE22W	VFIRE22	
154.2725	Fire Mutual Aid		VFIRE24	
154.2800 base/mobile	Fire Mutual Aid	VFIRE21W	VFIRE21	Not available in Puerto Rico and the U.S. Virgin Islands.
154.2875			VFIRE25	
154.2950 mobile	Fire Mutual Aid	VFIRE23W	VFIRE23	
154.3025			VFIRE26	
155.3400 base/mobile	EMS Mutual Aid	VMED28W	VMED28	May be designated for EMS Mutual Aid.
155.3475			VMED29	May be designated for EMS Mutual Aid.
155.4750 base/mobile	Law Enforcement Mutual Aid	VLAW31W	VLAW31	
155.4825	Law Enforcement Mutual Aid		VLAW32	

Rules for use of these channels are contained in 47 CFR 90.20 and NTIA Manual Section 4.3.11 & 7.3.6.

See also "Non-Federal VHF National Interoperability Channels" and "Non-Federal VHF Inland Interoperability Channels" on page 26 - 28 of this document.

UHF MED (Medical, EMS) Channels

Base & Mobile TX	Mobile TX	Channel Name	Bandwidth
462.950	467.950	MED–9 *	W,N,U
462.95625	467.95625	MED–91 *	U
462.9625	467.9625	MED–92 *	N,U
462.96875	467.96875	MED–93 *	U
462.975	467.975	MED–10 *	W,N,U
462.98125	467.98125	MED–101 *	U
462.9875	467.9875	MED–102 *	N,U
462.99375	467.99375	MED–103 *	U
* Used primarily for dispatch; may be used for mutual aid. 47CFR90.20(d)(65).			
463.000	468.000	MED–1	W,N,U
463.00625	468.00625	MED–11	U
463.0125	468.0125	MED–12	N,U
463.01875	468.01875	MED–13	U
463.025	468.025	MED–2	W,N,U
463.03125	468.03125	MED–21	U

Direct mode: receive & transmit on "Base & Mobile TX" freq.; Repeater mode: transmit on "Mobile TX" freq., receive on "Base & Mobile TX" freq. CTCSS as required by local plan. Bandwidth: W=wide, N=narrow, U=ultra-narrow (6.25 kHz). Add "D" to channel name when operating in "Direct" mode.

UHF MED (Medical, EMS) Channels			
Base & Mobile TX	Mobile TX	Channel Name	Bandwidth
463.0375	468.0375	MED–22	N,U
463.04375	468.04375	MED–23	U
463.050	468.050	MED–3	W,N,U
463.05625	468.05625	MED–31	U
463.0625	468.0625	MED–32	N,U
463.06875	468.06875	MED–33	U
463.075	468.075	MED–4	W,N,U
463.08125	468.08125	MED–41	U
463.0875	468.0875	MED–42	N,U
463.09375	468.09375	MED–43	U
463.100	468.100	MED–5	W,N,U
463.10625	468.10625	MED–51	U
463.1125	468.1125	MED–52	N,U
463.11875	468.11875	MED–53	U
463.125	468.125	MED–6	W,N,U

Direct mode: receive & transmit on "Base & Mobile TX" freq.; Repeater mode: transmit on "Mobile TX" freq., receive on "Base & Mobile TX" freq. CTCSS as required by local plan. Bandwidth: W=wide, N=narrow, U=ultra-narrow (6.25 kHz). Add "D" to channel name when operating in "Direct" mode.

UHF MED (Medical, EMS) Channels

Base & Mobile TX	Mobile TX	Channel Name	Bandwidth
463.13125	468.13125	MED–61	U
463.1375	468.1375	MED–62	N,U
463.14375	468.14375	MED–63	U
463.150	468.150	MED–7	W,N,U
463.15625	468.15625	MED–71	U
463.1625	468.1625	MED–72	N,U
463.16875	468.16875	MED–73	U
463.175	468.175	MED–8	W,N,U
463.18125	468.18125	MED–81	U
463.1875	468.1875	MED–82	N,U
463.19375	468.19375	MED–83	U

Direct mode: receive & transmit on "Base & Mobile TX" freq.; Repeater mode: transmit on "Mobile TX" freq., receive on "Base & Mobile TX" freq. CTCSS as required by local plan. Bandwidth: W=wide, N=narrow, U=ultra-narrow (6.25 kHz). Add "D" to channel name when operating in "Direct" mode.

The 25 Cities Project Federal Interoperability Channels

The 25 Cities Project Federal Interoperability Channels were developed through the Department of Justice "25 Cities" project to support local, state, federal, and tribal voice communications interoperability. Each metropolitan area has agreed upon policies and procedures regarding use of these channels. Most 25 Cities VHF channels are accessible by non-VHF users via permanent or ad hoc patching capabilities. All agencies interested in using these frequencies, who are not currently participating in the 25 Cities effort, should contact the local FBI Radio Manager prior to programming any equipment. For frequencies and programming details or other questions regarding the project, contact Rob Zanger, U.S. Department of Justice, Wireless Management Office at 202.598.2000 or robert.m.zanger@usdoj.gov.

Information as of November 17, 2010.

CITY	CHANNEL NAME
ATLANTA	ATL FIO (VHF P25 Voted System)
BALTIMORE	BAFIOLE3 (VHF P25 Voted System)
BOSTON	BPD FIO (VHF Voted System - Analog)
CHICAGO	CG-COM-N, CG-COM-C, CG-COM-S (VHF P25 Multicast Voted System)
	CG-TAC-N, CG-TAC-C, CG-TAC-S (VHF P25 Multicast Voted System)
DALLAS	DFW EAST (VHF P25 Voted System)
	DFW WEST (VHF P25 Voted System)
(continued)	

The 25 Cities Project Federal Interoperability Channels	
CITY	**CHANNEL NAME**
DENVER	DEN IO-N, DEN IO-E, DEN IO-C, DEN IO-S, DEN IO-W (VHF P25 Multicast Voted System)
EL PASO	EP FIO-W, EP FIO-E (VHF P25 Multicast Voted System)
HAMPTON ROADS – NORFOLK	HRN FIO (VHF P25 Voted System)
HARTFORD, CT	CFedcom-N, CFedcom-S (VHF P25 Multicast Voted System)
HONOLULU	HNL FIO (VHF P25 Stand Alone 125 watt repeater)
	HNL FIO2 (VHF P25 Stand Alone 125 watt repeater)
	LE 4 (VHF P25 Transportable 125 watt repeater)
	HNL FIRE (VHF Voted System – Analog)
HOUSTON	HOU CMD (VHF P25 Voted System)
	HOU PAT (VHF P25 Voted System)
JACKSONVILLE	JAX FIO (VHF P25 Voted System)
(continued)	

The 25 Cities Project Federal Interoperability Channels

CITY	CHANNEL NAME
LOS ANGELES	LA FIO1 (VHF P25 Voted System)
	LA FIO2 (VHF P25 Voted System)
	LA FIO3 (VHF P25 Voted System)
MIAMI	MIA FIO (VHF P25 Voted System)
MINNEAPOLIS/ST PAUL	FEDCOM-MP, FEDCOM-SP (VHF P25 Multicast Voted System)
NEW ORLEANS	NOLA FIO (VHF P25 Voted System)
NEW YORK	NYC FIO (NYC), NYC FIO-N (Orange-Putnam), NYC FIO-E (Suffolk), NYC FIO-S (Central NJ) (VHF P25 Multicast Voted system)
	NYC FIO2 (VHF P25 Voted System)
ORLANDO	ORL FIO (VHF P25 Voted System
PHILADELPHIA	PH FIO (VHF P25 Voted System)
(continued)	

The 25 Cities Project Federal Interoperability Channels	
CITY	**CHANNEL NAME**
ST LOUIS	STL CALL (VHF P25 Voted System)
	8CALL90(800 MHz Simulcast Voted Repeater System)
	STL TAC (VHF P25 Voted System)
	8TAC91 (800 MHz Simulcast Voted Repeater System)
	All of the above repeaters can be networked together.
(continued)	

The 25 Cities Project Federal Interoperability Channels	
CITY	**CHANNEL NAME**
SAN FRANCISCO	SF MA U-A (UHF Stand Alone 125 watt repeater - Analog)
	SF MA V-A (VHF Stand Alone 125 watt repeater - Analog)
	CLEMARS 7 (LLAW1) (Low Band repeater)
	SF MA T-A (UHF-T Band Stand Alone 125 watt repeater - Analog)
	8TAC94 (800 MHz Stand Alone 125 watt repeater- Analog)
	SF FED-V (VHF P25 Stand Alone 125 watt repeater)
	SF FED-U (UHF P25 Stand Alone 125 watt repeater)
	All of the above repeaters can be networked together.
	SF FED-ED, SF FED-ES, SF FED-ET, SF FED-EW (VHF P25 Multicast Voted System)
TAMPA	TAM FIO (VHF P25 Voted System)
WASHINGTON DC	DC IO-1 (VHF P25 Voted System)
	DCIO2LE2 (VHF P25 Voted System)

NOAA Weather Radio "All Hazards" Broadcasts

NWR broadcasts National Weather Service (NWS) warnings, watches, forecasts and other non-weather related hazard information 24 hours a day. Channels WX1-WX7 are used in the US & Canada; channels WX8-WX9 are used for Canada Marine Weather broadcasts in some areas. These channels should be programmed as wideband FM (16K0F3E) RECEIVE ONLY. Some radio manufacturers number the US weather channels in the order they came into use, others number them in frequency order. For programming in land-mobile radios, frequency order is recommended.

Weather Radio Broadcasts – Receive Only						
(WX1-WX7 US & Canada; WX8-WX9 Canada Marine Weather)						
WX1	WX2	WX3	WX4	WX5	WX6	WX7
162.400	162.425	162.450	162.475	162.500	162.525	162.550

Marine 21B	Marine 83B
WX8	WX9
161.650	161.775

– 49 –

COMMON COMMUNICATIONS REFERENCES
Operations Center Telephone Numbers

DHS Main Number 202-282-8000
NOC Senior Watch Officer 202-282-8101

FCC Federal Communications Commission
Communications and Crisis Management Center
(CCMC) e-mail comm-ctr@fcc.gov

FEMA Federal Emergency Management Agency,
National Response Coordination Center (NRCC) 202-646-2828
(general number for all ESFs – see next page) .... FEMA-NRCC@dhs.gov

NCS National Communications System
NCC Radio Room/SHARES HF Radio 703-235-5080
Operations Center / NCC Watch 703-235-5080

ARC American National Red Cross
24-hr Disaster Operations Center 800-526-3571, 202-303-5555

ARRL American Radio Relay League .... emergency@arrl.org
Main Number .... 860-594-0200 -0259 fax
Emergency Preparedness & Response Manager 860-594-0222
Radio Station W1AW 860-594-0268

General number: .... 202-418-1122, -2813 FAX

Emergency Support Functions (ESF)

ESF #1: Transportation	ESF #9: Urban Search & Rescue
ESF #2: Communications	ESF #10: Oil & Hazardous Materials Response
ESF #3: Public Works and Engineering	ESF #11: Agriculture and Natural Resources
ESF #4: Firefighting	ESF #12: Energy
ESF #5: Emergency Management	ESF #13: Public Safety and Security
ESF #6: Mass Care, Housing, and Human Services	ESF #14: Long-Term Community Recovery
ESF #7: Resource Support	ESF #15: External Affairs
ESF #8: Public Health and Medical Services	**Telephone number for all ESFs 202-646-2828**

FEMA Regions - States and Territories

Region I: CT, MA, ME, NH, RI, VT - 1-617-956-7506 or 1-877-336-2734

Region II: NJ, NY, Puerto Rico and the US Virgin Islands
NJ and NY: 1-212-680-3600
PR and USVI: 1-787-296-3500

Region III: DC, DE, MD, PA, VA, WV - 1-215-931-5500

Region IV: AL, FL, GA, KY, MS, NC, SC, TN - 1-770-220-5200

Region V: IL, IN, MI, MN, OH, WI - 1-312-408-5500

Region VI: AR, LA, NM, OK, TX - 1-940-898-5399

Region VII: IA, KS, MO, NE - 1-816-283-7061

Region VIII: CO, MT, ND, SD, UT, WY - 1-303-235-4800

Region IX: AZ, CA, Guam (GU), HI, NV, CNMI, RMI, FSM, American Samoa (AS)
1-510-627-7100

Region X: AK, ID, OR, WA - 1-425-487-4600

FEMA Headquarters, Washington DC: 1-202-646-2500

FEMA Disaster Assistance: 1-800-621-FEMA (3362)

U.S. Coast Guard Rescue Coordination Centers

24 hour Regional Contacts for Emergencies
Last Modified 4/29/2009

RCC	Location	Phone Number
Atlantic Area SAR Coordinator	*Portsmouth, VA*	*(757)398-6700*
RCC Boston	Boston, MA	(617)223-8555
RCC Norfolk	Portsmouth, VA	(757)398-6231
RCC Miami	Miami, FL	(305)415-6800
RSC San Juan	San Juan, PR	(787)289-2042
RCC New Orleans	New Orleans, LA	(504)589-6225
RCC Cleveland	Cleveland, OH	(216)902-6117
Pacific SAR Coordinator	*Alameda, CA*	*(510)437-3700*
RCC Alameda	Alameda, CA	(510)437-3700
RCC Seattle	Seattle, WA	(206)220-7001
RCC Honolulu	Honolulu, HI	(808) 535-3333
Sector Guam	Apra Harbor, GU	(671)355-4824
RCC Juneau	Juneau, Alaska	(907)463-2000

CTCSS Tones and Codes

Freq. (Hz)	Motorola Code	NIFC & CA Fire *	Freq. (Hz)	Motorola Code	NIFC & CA Fire *
67.0	XZ		136.5	4Z	4
69.3**	WZ		141.3	4A	13
71.9	XA		146.2	4B	5
74.4	WA		151.4	5Z	14
77.0	XB		156.7	5A	6
79.7	WB		162.2	5B	15
82.5	YZ		167.9	6Z	7
85.4	YA		173.8	6A	
88.5	YB		179.9	6B	
91.5	ZZ		186.2	7Z	
94.8	ZA		192.8	7A	16
97.4	ZB		203.5	M1	
100.0	1Z	9	206.5	8Z	
103.5	1A	8	210.7	M2	
107.2	1B	10	218.1	M3	
110.9	2Z	1	225.7	M4	
114.8	2A	11	229.1	9Z	
118.8	2B		233.6	M5	
123.0	3Z	2	241.8	M6	
127.3	3A	12	250.3	M7	
131.8	3B	3	254.1	0Z	

* California FIRESCOPE tone list, used by NIFC and CA fire agencies
  Ref. http://www.firescope.org/macs-docs/MACS-441-1.pdf
** 69.4 in some radios

DCS Codes

Normal	Inverted	Nor.	Inv.	Nor.	Inv.	Nor.	Inv.
023	047	155	731	325	526	516	432
025	244	156	265	331	465	523	246
026	464	162	503	332	455	526	325
031	627	165	251	343	532	532	343
036	171	172	036	346	612	546	132
043	445	174	074	351	243	565	703
047	023	205	263	364	131	606	631
051	032	212	356	365	125	612	346
053	452	223	134	371	734	624	632
054	413	225	122	411	226	627	031
065	271	226	411	412	143	631	606
071	306	243	351	413	054	632	624
072	245	244	026	423	315	654	743
073	506	245	072	431	723	662	466
074	174	246	523	432	516	664	311
114	712	251	165	445	043	703	565
115	152	252	462	446	255	712	114
116	754	255	446	452	053	723	431
122	225	261	732	454	266	731	155
125	365	263	205	455	332	732	261
131	364	265	156	462	252	734	371
132	546	266	454	464	026	743	654
134	223	271	065	465	331	754	116
143	412	274	145	466	662		
145	274	306	071	503	162		
152	115	311	664	506	073		
032	051	315	423				

P25 Digital Codes

NAC – Network Access Codes

$293	default NAC
$F7E	receiver will unsquelch with any incoming NAC
$F7F	a repeater with this NAC will allow incoming signals to be repeated with the NAC intact

TGID – Talkgroup ID

$0001	default
$0000	no-one, talkgroup with no users – used for individual call
$FFFF	talkgroup which includes everyone

Unit ID

$000000	no-one – never associated with a radio unit
$000001-$98767F	for general use
$989680-$FFFFFE	for talkgroup use or other special purposes
$FFFFFF	designates everyone – used when implementing a group call with a TGID3

RS-232 Connectors (DB25 and DB9)

"Front" refers to the ends with the pins; "rear" refers to the end with the cable. The following is a view of the pins, looking at the front of the female connector (rear of male):

same for DB25, except top pins 13 - 1, bottom 25 - 14 (left to right)

DB9	DB25	Signal
1	8	Carrier Detect
2	3	Receive Data
3	2	Transmit Data*
4	20	Data Terminal Ready*
5	1,7	Ground **
6	6	Data Set Ready
7	4	Request to Send*
8	5	Clear to Send
9	22	Ring Indicator
* An output from the computer to the outside world.		
** On the DB25, 1 is the protective ground, 7 is the signal ground.		

RJ-45 Wiring

		T568A (less common)		T568B (more common)	
Pin	Pair	Color	Name	Color	Name
1	2	white/ green	RecvData+	white/orange	TxData +
2	2	green	RecvData-	orange	TxData -
3	3	white/orange	TxData +	white/green	RecvData+
4	1	blue		blue	
5	1	white/blue		white/blue	
6	3	orange	TxData -	green	RecvData-
7	4	white/brown		white/brown	
8	4	brown		brown	

Note that the odd pin numbers are always the white-with-stripe color.

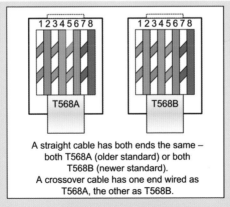

A straight cable has both ends the same –
both T568A (older standard) or both
T568B (newer standard).
A crossover cable has one end wired as
T568A, the other as T568B.

Telephone Connectors

Pin numbers are from left to right, holding the plug with the contacts up and looking at the side that does not have the spring clip. "T" and "R" indicate "Tip" and "Ring".

Pin	RJ25	RJ14	RJ11
1	T3		
2	T2	T2	
3	R1	R1	R1
4	T1	T1	T1
5	R2	R2	
6	R3		

Circuit	Twisted-Pair Colors	25-Pair Colors	Solid Colors
T1	White/Blue	White/Blue	Green
R1	Blue	Blue/White	Red
T2	White/Orange	White/Orange	Black
R2	Orange	Orange/White	Yellow
T3	White/Green	White/Green	White
R3	Green	Green/White	Blue
T4	White/Brown	White/Brown	Orange
R4	Brown	Brown/White	Brown

Telephone Block Wiring

Tip, Ring	Tip Color (reverse for Ring)	50 Pin Position	66 or 110 Block Position
1	White/Blue	26,1	1,2
2	White/Orange	27,2	3,4
3	White/Green	28,3	5,6
4	White/Brown	29,4	7,8
5	White/Slate	30,5	9,10
6	Red/Blue	31,6	11,12
7	Red/Orange	32,7	13,14
8	Red/Green	33,8	15,16
9	Red/Brown	34,9	17,18
10	Red/Slate	35,10	19,20
11	Black/Blue	36,11	21,22
12	Black/Orange	37,12	23,24
13	Black/Green	38,13	25,26
14	Black/Brown	39,14	27,28
15	Black/Slate	40,15	29,30
16	Yellow/Blue	41,16	31,32
17	Yellow/Orange	42,17	33,34
18	Yellow/Green	43,18	35,36
19	Yellow/Brown	44,19	37,38
20	Yellow/Slate	45,20	39,40
21	Violet/Blue	46,21	41,42
22	Violet/Orange	47,22	43,44
23	Violet/Green	48,23	45,46
24	Violet/Brown	49,24	47,48
25	Violet/Slate	50,25	49,50

Telephone Keypad Letters

1:(QZ)	2:ABC	3:DEF
4:GHI	5:JKL	6:MNO
7:P(Q)RS	8:TUV	9:WXY(Z)
*	0	#

DSN Area Codes

Defense Switched Network - Global Operator – 1-719-567-1110 (DSN 312-560-1110)

312 - CONUS   313 – Caribbean
314 - Europe   315 - Pacific
317 - Alaska   318 - Southwest Asia
319 - Canada

DSN Directory - Global http://www.disa.mil/dsn/directory/global.html

Cellular Telephone Emergency Response

Some cellular telephone companies have transportable cell sites (Cellular On Wheels – COWs, Cellular on Light Trucks – COLTs, etc.) that can be deployed during disasters, emergencies, and special events. Local jurisdictions are encouraged to coordinate with their established service provider representatives for local events; however, the U.S. Department of Homeland Security – National Communications System National Coordinating Center will assist jurisdictions with referrals to corporate level contacts for wireless/wireline service provider representatives if needed.

The NCS-NCC 24x7 Watch can be reached at 1-703-235-5080 or e-mail NCS@dhs.gov

Satellite Phone Dialing Instructions

From a US Landline
(helpful directions for someone to call you back)

To an Iridium phone directly as an International Call
> 011 + 8816xxxxxxx (Iridium Phone Number)

To an M4 phone directly as an International Call
> 011 + 870 + 76xxxxxxx (Mobile Number)

Iridium PIN (default) is 1111
(enter when powering-on the Iridium Subscriber Unit)

From an M4: [Note - Cannot call Toll-Free numbers]

To a US Phone number:
> 00 + 1 + (10-digit US phone number)

To an Iridium phone directly
> 00 + 8816xxxxxxx (Iridium Phone Number)

To an M4 phone directly
> 00 + 870 + 76xxxxxxx (Mobile Number)

From an Iridium provisioned commercially

To a US Phone number
> 00 + 1 + xxx.xxx.xxxx (US phone number)

To an Iridium phone directly
> 00 + 8816xxxxxxx (Iridium Phone Number)

To an M4 phone directly
> 00 + 870 + 76xxxxxxx (Mobile Number)

Test call - no airtime charge: 00 + 1 + 480.752.5105

From an Iridium provisioned by DOD

ISU (Iridium Subscriber Unit) to DSN
    00 + 696 + (DSN Area Code) + (DSN 7-digit number)

ISU to U.S. Domestic
    00 + 697 + (U.S. Area Code) + (7-digit US number)

ISU to International Long Distance (ILD)
    00 + 698 + (Country Code) + ("National Destination Code" or
    "City Code") + (Subscriber Number)

ISU to INMARSAT
    00 + 698 + 870 + (INMARSAT subscriber number)

ISU to Local Hawaii
    00 + 699 + (7-digit local commercial number)
    1-800 toll-free 00 + 699 + 1+ 800 + (7-digits)

ISU to ISU, handset-to-handset
    00 + (12-digit ISU subscriber number, e.g., 8816 763-xxxxx)

INMARSAT Country Code

All INMARSAT satellite telephones now use country code 870.
The Ocean Region Codes were discontinued January 1, 2009:
    871   Atlantic Ocean Region – East [AOR-East]
    872   Pacific Ocean Region [POR]
    873   Indian Ocean Region [IOR]
    874   Atlantic Ocean Region – West [AOR-West]
Inmarsat Customer Care Helpline - international direct dialing from USA
to London, United Kingdom: 011 44 20 7728 1030

INMARSAT-M Service Codes	
00	Automatic Calls
11	International Operator
12	International Information
13	National Operator
14	National Information
17	Telephone Call Booking
20	Access to a Maritime PAD
23	Abbreviated Dialing
24	Post FAX
31	Maritime Enquiries
32	Medical Advice
33	Technical Assistance
34	Person-to-Person Call
35	Collect Call
36	Credit Card Call
37	Time and Duration
38	Medical Assistance
39	Maritime Assistance
41	Meteorological Reports
42	Navigational Hazards and Warnings
43	Ship Position Reports
57	Retrieval of Mailbox Messages
6x	Administration, Specialized Use
70	Databases
91	Automatic Line Test
911	Emergency Calls
92	Commissioning Tests

Wireless Priority Service (WPS)

Authorized phones only; monthly and usage charges apply. http://wps.ncs.gov/

Dial *272 + destination number [send]

GETS - Govt. Emergency Telecomm. Service

User Assistance: 1-800-818-GETS, 1-703-818-GETS

http://www.ncs.gov   GETS test #: 1-703-818-3924

GETS call from a commercial phone:	
1-710-NCS-GETS (1-710-627-4387)	1-888-288-GETS (ATT)
1-800-900-GETS (MCI/Verizon)	1-800-257-8373 (Sprint)

Optional: specify long-distance carrier
       1010+288 (ATT) 1-710-NCS-GETS
       1010+222 (Verizon) 1-710-NCS-GETS
       1010+333 (Sprint) 1-710-NCS-GETS	

Listen for tone; enter PIN	
At prompt, enter 10-digit destination number	

GETS call from a rotary or pay phone:		
Get outside line, listen for dial tone		
Optional: specify long-distance carrier		
ATT: 1010+288	Verizon: 1010+222	Sprint: 1010+333
Dial 1-710-NCS-GETS (627-4387)		
Wait for GETS operator		
Give your PIN and 10-digit destination number		

Text Messaging	
Selected US & Canadian Cellular Text Messaging Carriers	
"number" is the 10-digit telephone number	
Alltel	number@alltelmessage.com or number@message.alltel.com
AT&T	number@txt.att.net or number@mms.att.net
Bell Canada (Mobility) - phone - blackberry	number@txt.bell.ca number@txt.bellmobility.ca
Centennial Wireless	number@cwemail.com
Cellular South	number@csouth1.com
Cincinnati Bell	number@gocbw.com
Metro PCS	number@mymetropcs.com or number@metropcs.sms.us
Nextel	number@messaging.nextel.com
Omnipoint	number@omnipointpcs.com
Qwest	number@qwestmp.com
Southernlinc	number @page.southernlinc.com
Sprint	number@messaging.sprintpcs.com
Suncom	number@tms.suncom.com
T-Mobile	number@tmomail.net
TracFone	number@mmst5.tracfone.com
Continued	

Text Messaging (continued)	
Telus	number@msg.telus.com
U.S. Cellular	number@email.uscc.net
Verizon	number@vtext.com
Virgin Mobile	number@messaging.sprintpcs.com

Alaska	
Alaska Communications Systems (ACS)	number@msg.acsalaska.com
General Communications Inc. (GCI)	number@mobile.gci.net

Puerto Rico	
Centennial Wireless	number@cwemail.com
Claro	number@vtexto.com
TracFone	number@mmst5.tracfone.com

U.S. Virgin Islands	
Centennial Wireless	number@cwemail.com
TracFone	number@mmst5.tracfone.com

See http://en.wikipedia.org/wiki/List_of_SMS_gateways for more.

Line-of-Sight Formulas

Visual Line-of-Sight

Approximate distance in miles = $1.33 \times \sqrt{\text{(height in feet)}}$

Radio Line-of-Sight

$D = \sqrt{(2Hr)} + \sqrt{(2Ht)}$

Where:
D = approximate distance (range) to radio horizon in miles
Hr = height of receive antenna in feet
Ht = height of transmit antenna in feet

Range (miles)	Tx Ant. Height (ft)	Rx Ant. Height (ft)
8	10	5.5
10	20	5.5
11	30	5.5
12	40	5.5
13	50	5.5
16	75	5.5
17	100	5.5

Range (miles)	Tx Ant. Height (ft)	Rx Ant. Height (ft)
21	150	5.5
23	200	5.5
28	300	5.5
32	400	5.5
35	500	5.5
42	750	5.5
48	1000	5.5

COMMONLY USED FREQUENCIES
Aviation Frequencies

121.5 Emergency & Distress

122.9 SAR Secondary and Training

123.1 SAR

122.925 – for use only for communications with or between aircraft when coordinating natural resources programs of Federal or State natural resources agencies, including forestry management and fire suppression, fish and game management and protection and environmental monitoring and protection.

Typical Uses	Fixed Wing	Rotary Wing
Air-to-Air	122.750 F	
	122.850 M	122.850 M
	122.925 M	122.925 M
	122.975 U	122.975 U
		123.025 A
	123.075 U	123.075 U
Air-to-Ground	122.850 M	122.850 M
	122.925 M	122.925 M
	122.975 U	122.975 U
		123.025 A
	123.075 U	123.075 U

A – Helicopter air-to-air, air traffic control operations.

F – Fixed-wing air-to-air.

M – Multicom.

U – Unicom.

Ask FAA/FCC for emergency use of 123.3 or 123.5 (flight training).

VHF Marine Channel Listing

This chart summarizes a portion of the FCC rules -- 47 CFR 80.371(c) and 80.373(f)

Type of Message	Appropriate Channels *
DISTRESS SAFETY AND CALLING - Use this channel to get the attention of another station (calling) or in emergencies (distress and safety).	16
INTERSHIP SAFETY - Use this channel for ship-to-ship safety messages and for search and rescue messages to ships and aircraft of the Coast Guard.	6
COAST GUARD LIAISON - Use this channel to talk to the Coast Guard (but first make contact on Channel 16).	22A
COAST GUARD - These channels are Coast Guard working channels, not available to commercial or non-commercial vessels for normal use.	21A, 23A, 81A, 83A
U.S. Government - Environmental protection operations.	81A
U.S. Government - This channel is a working channel for U.S. Government vessels and U.S. Government coast stations only.	82A
NONCOMMERCIAL - Working channels for voluntary boats. Messages must be about the needs of the ship. Typical uses include fishing reports, rendezvous, scheduling repairs and berthing information. Use Channels 67 and 72 only for ship-to-ship messages.	9[6], 67[9],68, 69, 71[8], 72, 78A, 79A[4], 80[4]

Type of Message	Appropriate Channels *
COMMERCIAL - Working channels for working ships only. Messages must be about business or the needs of the ship. Use channels 8, 67, 72 and 88A only for ship-to-ship messages.	1[5], 7A, 8, 9, 10, 11, 18A, 19A, 63[5], 67[7], 79A, 80A, 88A[1]
PUBLIC CORRESPONDENCE (MARINE OPERATOR) - Use these channels to call the marine operator at a public coast station. By contacting a public coast station, you can make and receive calls from telephones on shore. Except for distress calls, public coast stations usually charge for this service.	24, 25, 26, 27, 28, 84, 85, 86
PORT OPERATIONS - These channels are used in directing the movement of ships in or near ports, locks or waterways. Messages must be about the operational handling movement and safety of ships. In certain major ports, Channels 11, 12 and 14 are not available for general port operations messages. Use channel 20 only for ship-to-coast messages. Channel 77 is limited to intership communications to and from pilots.	1[5], 5[3], 12, 14, 20, 63[5], 65, 66, 73, 74, 75[10], 76[10], 77
NAVIGATIONAL - (Also known as the bridge-to-bridge channel.) This channel is available to all ships. Messages must be about ship navigation, for example, passing or meeting other ships. You must keep your messages short. Your power output must not be more than one watt. This is also the main working channel at most locks and drawbridges.	13, 67

Type of Message	Appropriate Channels *
MARITIME CONTROL - This channel may be used to talk to ships and coast stations operated by state or local governments. Messages must pertain to regulation and control, boating activities, or assistance to ships.	17
DIGITAL SELECTIVE CALLING - Use this channel for distress and safety calling and for general purpose calling using only digital selective calling techniques.	70
WEATHER - On these channels you may receive weather broadcasts of the National Oceanic and Atmospheric Administration. These channels are only for receiving. You cannot transmit on them.	WX-1 through WX-7

Footnotes

1. Not available in the Great Lakes, St. Lawrence Seaway, or the Puget Sound and the Strait of Juan de Fuca and its approaches.

2. Only for use In the Great Lakes, St Lawrence Seaway, and Puget Sound and the Strait of Juan de Fuca and its approaches.

3. Available only in the Houston and New Orleans areas.

4. Available only in the Great Lakes.

5. Available only in the New Orleans area.

6. Available for intership, ship, and coast general purpose calling by noncommercial ships.

7. Available only In the Puget Sound and the Strait of Juan de Fuca.

Type of Message	Appropriate Channels *
8. Available for port operations communications only within the U.S. Coast Guard designated VTS radio protection area of Seattle (Puget Sound). Normal output must not exceed 1 watt.	
9. Available for navigational communications only in the Mississippi River/ Southwest Pass/Gulf outlet area.	
10. Available for navigation-related port operations or ship movement only. Output power limited to 1 watt.	
* "A" indicates simplex use of the ship station transmit frequency of an international duplex channel. Used in U.S. waters only.	
December 21, 2010 Adapted from http://wireless.fcc.gov/services/index.htm?job=service_bandplan&id=ship_stations	

Shipboard repeaters: 457.525 457.550 457.575 457.600 MHz
Inputs are +10.225 MHz (foreign vessels may use +10.0 MHz offset – not permitted in U.S. waters).

Maritime freqs. assignable to aircraft:
      (HF) 2.738 2.830 3.023 4.125 5.680 MHz
      (VHF) channels 6 8 9 16 18A 22A 67 68 72 & 88A
      See 47CFR80.379 for restrictions.

VHF Marine Channels & Frequencies

Source: http://www.navcen.uscg.gov/?pageName=mtVhf

Channel Number *	Ship Transmit MHz	Ship Receive MHz	Use
01A	156.050	156.050	Port Operations and Commercial, VTS. Available only in New Orleans/Lower Mississippi area
05A	156.250	156.250	Port Operations or VTS in the Houston, New Orleans and Seattle areas
6	156.300	156.300	Intership Safety
07A	156.350	156.350	Commercial
8	156.400	156.400	Commercial (Intership only)
9	156.450	156.450	Boater Calling. Commercial and Non-Commercial
10	156.500	156.500	Commercial
11	156.550	156.550	Commercial. VTS in selected areas
12	156.600	156.600	Port Operations. VTS in selected areas

* "A" indicates simplex use of the ship station transmit frequency of an international duplex channel. Used in U.S. waters only.

Channel Number *	Ship Transmit MHz	Ship Receive MHz	Use
13	156.650	156.650	Intership Navigation Safety (Bridge-to-bridge). Ships >20m length maintain a listening watch on this channel in US waters.
14	156.700	156.700	Port Operations. VTS in selected areas.
15	--	156.750	Environmental (Receive only). Used by Class C EPIRBs.
16	156.800	156.800	International Distress, Safety and Calling. Ships required to carry radio, USCG, and most coast stations maintain a listening watch on this channel.
17	156.850	156.850	State & Local Government Maritime Control
18A	156.900	156.900	Commercial
19A	156.950	156.950	Commercial
20	157.000	161.600	Port Operations (duplex)
20A	157.000	157.000	Port Operations
21A	157.050	157.050	U.S. Coast Guard only

*"A" indicates simplex use of the ship station transmit frequency of an international duplex channel. Used in U.S. waters only.

Channel Number *	Ship Transmit MHz	Ship Receive MHz	Use
22A	157.100	157.100	Coast Guard Liaison and Maritime Safety Information Broadcasts. Broadcasts announced on channel 16.
23A	157.150	157.150	U.S. Coast Guard only
24	157.200	161.800	Public Correspondence (Marine Operator)
25	157.250	161.850	Public Correspondence (Marine Operator)
26	157.300	161.900	Public Correspondence (Marine Operator)
27	157.350	161.950	Public Correspondence (Marine Operator)
28	157.400	162.000	Public Correspondence (Marine Operator)
63A	156.175	156.175	Port Operations and Commercial, VTS. Available only in New Orleans/Lower Mississippi area.
65A	156.275	156.275	Port Operations
66A	156.325	156.325	Port Operations
67	156.375	156.375	Commercial. Used for bridge-to-bridge communications in lower Mississippi River. Intership only.

* "A" indicates simplex use of the ship station transmit frequency of an international duplex channel. Used in U.S. waters only.

Channel Number *	Ship Transmit MHz	Ship Receive MHz	Use
68	156.425	156.425	Non-Commercial
69	156.475	156.475	Non-Commercial
70	156.525	156.525	Digital Selective Calling (voice communications not allowed)
71	156.575	156.575	Non-Commercial
72	156.625	156.625	Non-Commercial (intership only)
73	156.675	156.675	Port Operations
74	156.725	156.725	Port Operations
77	156.875	156.875	Port Operations (intership only)
78A	156.925	156.925	Non-Commercial
79A	156.975	156.975	Commercial. Non-Commercial in Great Lakes only
80A	157.025	157.025	Commercial. Non-Commercial in Great Lakes only
81A	157.075	157.075	U.S. Government only - Environmental protection operations.
82A	157.125	157.125	U.S. Government only

* "A" indicates simplex use of the ship station transmit frequency of an international duplex channel. Used in U.S. waters only.

Channel Number *	Ship Transmit MHz	Ship Receive MHz	Use
83A	157.175	157.175	U.S. Coast Guard only
84	157.225	161.825	Public Correspondence (Marine Operator)
85	157.275	161.875	Public Correspondence (Marine Operator)
86	157.325	161.925	Public Correspondence (Marine Operator)
87A	157.375	157.375	Public Correspondence (Marine Operator)
88A	157.425	157.425	Commercial, intership only.
AIS 1	161.975	161.975	Automatic Identification System (AIS)
AIS 2	162.025	162.025	Automatic Identification System (AIS)
* "A" indicates simplex use of the ship station transmit frequency of an international duplex channel. Used in U.S. waters only.			

Multi-Use Radio Service (MURS)

151.820 MHz

151.880 MHz

151.940 MHz

154.570 MHz (shared with business band)

154.600 MHz (shared with business band)

Maximum power output 2 watts.

Narrowband on 151 MHz frequencies.
Narrowband or wideband on the 154 MHz frequencies.

External gain antennas may be used (must be no more than 60 feet above ground or 20 feet above the structure on which it is mounted).

Voice or data, except:

      no store-and-forward packet operation

      no continuous carrier operation

      no interconnection with the public switched network

      no use aboard aircraft in flight

Authorized emission types:

      A1D, A2B, A2D, A3E, F2B, F1D, F2D, F3E, G3E.

Personal or business use.

Equipment must be certificated per FCC rules Part 95, Subpart J.

No license required.

GMRS Frequencies
Repeater outputs (inputs are +5 MHz):

462.550 462.575 462.600 462.625 462.650 462.675* 462.700 462.725
* nationwide traveler's assistance; if CTCSS is required, try 141.3 Hz.

Simplex prohibited on repeater inputs.

Interstitial frequencies (simplex, not more than 5 watts):
462.5625 .5875 .6125 .6375 .6625 .6875 .7125 (shared with FRS)

FRS Frequencies
(Channels 1-14)

462.5625 /5875 /6125 /6375 /6625 /6875 /7125 (shared with GMRS)
467.5625 /5875 /6125 /6375 /6625 /6875 /7125

CB Frequencies

Ch	MHz	Ch	MHz	Ch	MHz	Ch	MHz	Ch	MHz
1	26.965	2	26.975	3	26.985	4	27.005	5	27.015
6	27.025	7	27.035	8	27.055	9	27.065	10	27.075
11	27.085	12	27.105	13	27.115	14	27.125	15	27.135
16	27.155	17	27.165	18	27.175	19	27.185	20	27.205
21	27.215	22	27.225	23	27.255	24	27.235	25	27.245
26	27.265	27	27.275	28	27.285	29	27.295	30	27.305
31	27.315	32	27.325	33	27.335	34	27.345	35	27.355
36	27.365	37	27.375	38	27.385	39	27.395	40	27.405
*	26.995	*	27.045	*	27.095	*	27.145	*	27.195
* Remote Control Channels									

Common Business Frequencies

IS=Special Industrial IB=Business

27.49	IB	Itinerant
35.04	IB	Itinerant
43.0400	IS	Itinerant
151.5050	IS	Itinerant
151.6250	IB	RED DOT Itinerant
151.9550	IB	PURPLE DOT
152.8700	IS	Itinerant
154.5700	IB	BLUE DOT (also MURS)
154.6000	IB	GREEN DOT (also MURS)
158.4000	IS	Itinerant
451.8000	IS	Itinerant
456.8000	IS	Itinerant
464.5000	IB	BROWN DOT Itinerant 35w.
464.5500	IB	YELLOW DOT Itinerant 35w.
467.7625	IB	J DOT
467.8125	IB	K DOT
467.8500	IB	SILVER STAR
467.8750	IB	GOLD STAR
467.9000	IB	RED STAR
467.9250	IB	BLUE STAR
469.5000	IB	Simplex or input to 464.500 if repeater. Itinerant 35 w. max
469.5500	IB	Simplex or input to 464.550 if repeater. Itinerant 35 w. max

Railroad Frequencies

160.215(ch.7)-161.565(ch.97), every 15 kHz
  Interstitial narrowband channels between ch. 2-97 are offset 7.5 kHz.

161.205 Railroad Police Mutual Aid
        (Wideband: channel 73; narrowband: channel 073)
 Ch. 2-6 are used in Canada only:
        159.810 159.930 160.050 160.185 160.200

452.325 / 457.325
452.375 / 457.375
452.425 / 457.425
452.475 / 457.475

452.775 / 457.775
452.825 / 457.825
452.875 / 452.875
452.900 / 457.900

452.8500
452.8375 - low power
452.8625 - low power
452.8875 - low power

(telemetry / remote control / remote indicator frequencies omitted)

SAR (Search And Rescue) Frequencies

Land SAR

Typical freqs. are: 155.160, .175, .205, .220, .235, .265, .280, or .295
If CTCSS is required try 127.3 Hz (3A).

Air SAR

3023, 5680, 8364 kHz upper sideband (lifeboat/survival craft),
4125 kHz upper sideband (distress/safety with ships and coast stations)
121.5 MHz emergency and distress
122.9 MHz SAR secondary & training
123.1 MHz SAR primary

Water SAR

156.300 (VHF Marine ch. 06) Safety and SAR
156.450 (VHF Marine ch. 09) Non-commercial supplementary calling
156.800 (VHF Marine ch. 16) DISTRESS and calling
156.850 (VHF Marine ch. 17) State & Local Government Maritime Control
157.100 (VHF Marine ch. 22A) Coast Guard Liaison

VHF Marine Channels

6, 9, 15, 16, 21A, 22A (USCG Liaison), 23A, 81A, 83A

USCG Auxiliary

138.475, 142.825, 143.475, 149.200, 150.700

USCG/DOD Joint SAR

345.0 MHz AM initial contact, 282.8 MHz AM working

Military SAR

40.50 wideband FM	US Army/USN SAR
138.450 AM, 138.750 AM	USAF SAR

TEXAS COUNTIES WHERE VTAC17/VTAC17D MAY BE USED

(see page 28)

Andrews	Donley	Kimble	Randall
Armstrong	Ector	King	Reagan
Bailey	Edwards	Kinney	Reeves
Borden	El Paso	Knox	Roberts
Brewster	Fisher	Lamb	Runnels
Briscoe	Floyd	Lipscomb	Schleicher
Callahan	Gaines	Loving	Scurry
Carson	Garza	Lubbock	Sherman
Castro	Glasscock	Lynn	Sterling
Childress	Gray	McCulloch	Stonewall
Cochran	Hale	Martin	Sutton
Coke	Hall	Menard	Swisher
Collingsworth	Hansford	Midland	Taylor
Concho	Hartley	Mitchell	Terrell
Cottle	Haskell	Moore	Terry
Crane	Hockley	Motley	Tom Green
Crockett	Howard	Nolan	Upton
Crosby	Hudspeth	Ochiltree	Val Verde
Culberson	Hutchinson	Oldham	Ward
Dallam	Irion	Parmer	Wheeler
Dawson	Jeff Davis	Pecos	Winkler
Deaf Smith	Jones	Potter	Yoakum
Dickens	Kent	Presidio	

NOTES

NOTES

NOTES

NOTES

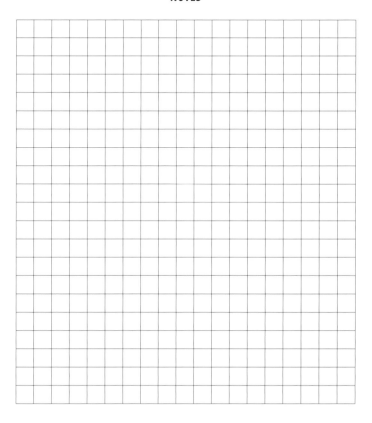

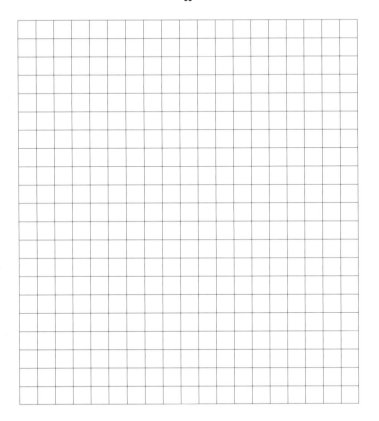

NOTES

NOTES

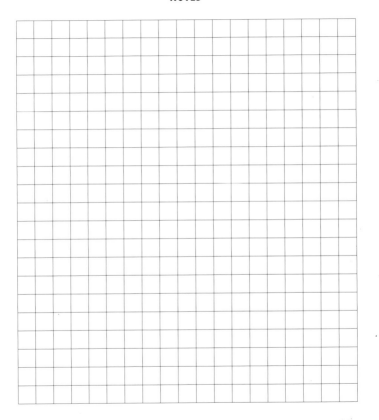

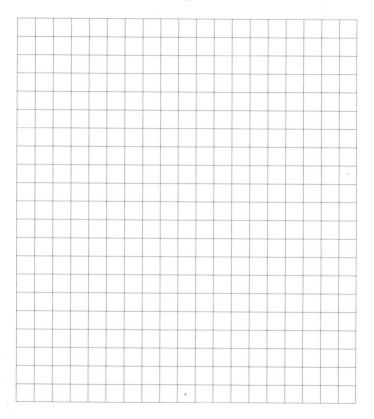